Pond Life

Library of Congress Number: 77-27243

1 2 3 4 5 6 7 8 9 0 82 81 80 79 78

Printed and bound in the United States of America.

Library of Congress Cataloging in Publication Data

Kirkpatrick, Rena K.
    Look at pond life.

    Includes index.
    SUMMARY: Easy-to-read text and illustrations
explore the plant and animal life in a pond.

    1. Pond ecology—Juvenile literature. [1. Pond
ecology. 2. Ecology] I. Milne, Annabel.
II. Stebbing, Peter. III. Title.
QH541.5.P63W5   574.5'2632   77-27243
ISBN 0-8393-0059-X lib. bdg.

Look At
POND LIFE

Words by Rena K. Kirkpatrick
Science Consultant

Pictures by Annabel Milne and Peter Stebbing

Raintree Childrens Books
Milwaukee • Toronto • Melbourne • London

willow
tree

alder
tree

common
reed

juncus
rush

water lily

water hen

water
plantain

The pond is like a community. There is food, water, and shelter for many animals. Plants also live in a pond. Look at the many plants and animals of the pond.

arrowhead

yellow
flag

When we stand on the bank, we
cannot see everything that lives in a
pond. The children are very quiet so
they do not disturb the water hen or
the frog.

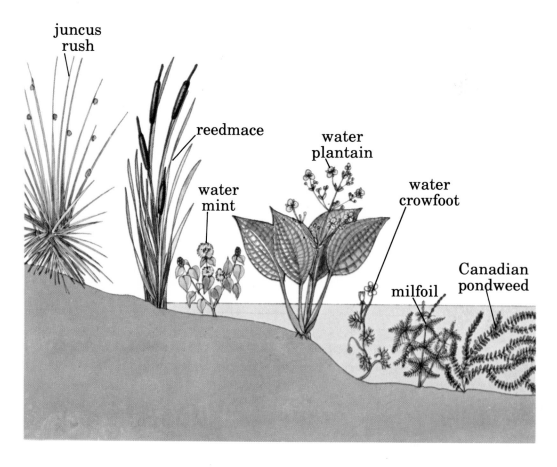

Let's look at some pond plants.
Some plants have roots in the bottom
of the pond. Some plants have
roots that float. Some grow only
underwater.

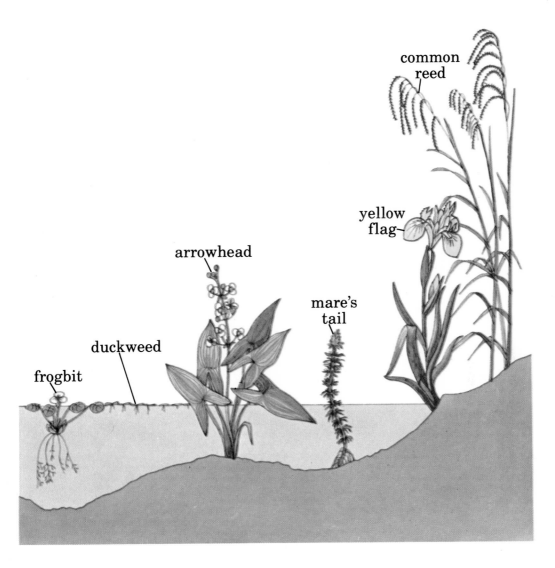

common
reed

yellow
flag

arrowhead

mare's
tail

duckweed

frogbit

Some water plants have pretty
flowers. Plants like the reeds and
rushes grow on the bank of the pond.
They need lots of water.

water hen

duck

drake

Many birds live on or near the pond. Some build their nests on the shore. Other birds build nests in the middle of the pond. The water hen feels safer in the middle of the pond.

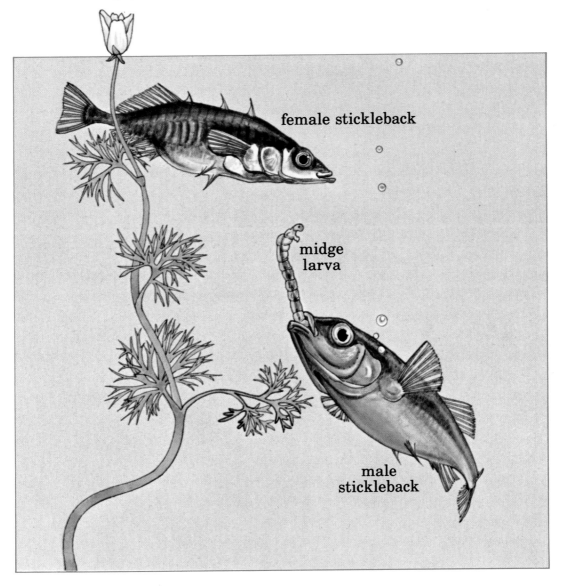

female stickleback

midge
larva

male
stickleback

Fish live in the water. Many
different kinds may live in ponds. The
stickleback is one fish that lives in a
pond. The male stickleback is brightly
colored. Fish eat plants from the water
and also insects.

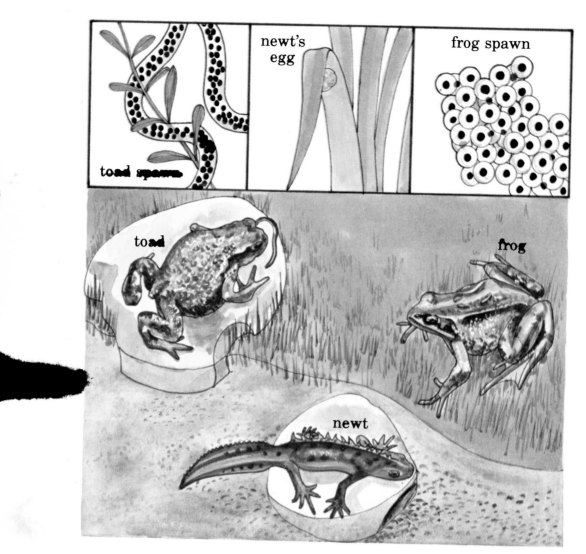

Frogs, toads, and newts lay eggs in the spring. Frogs and toads lay their eggs in the water near the shore. Newts lay their eggs in the grass on the shore. The young animals that hatch from the eggs live in the water.

pondweed

growing tadpoles

The girl has some frog eggs from the pond. They have hatched into tadpoles. Tadpoles have a tail. They have no legs. The young tadpoles eat water plants.

As the tadpoles get older they begin to grow legs. They then need raw meat to eat. If they were in the pond, they would be eating insects. When all four legs have grown, the children take them back to the pond.

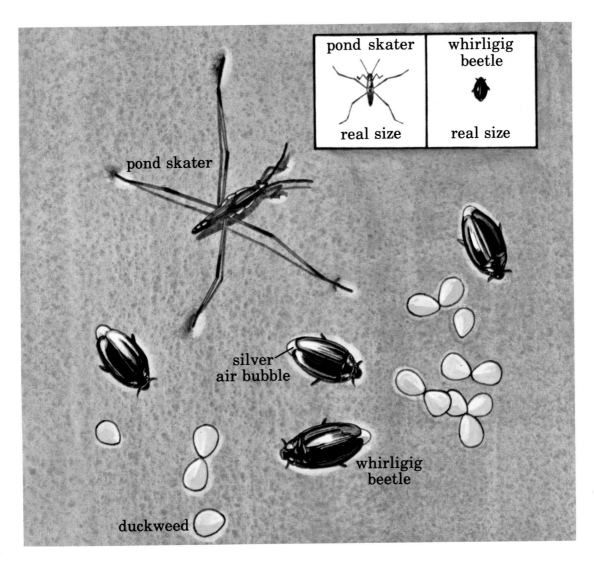

pond skater

whirligig beetle

real size

real size

pond skater

silver air bubble

whirligig beetle

duckweed

There are many insects near a pond. They are food for fish, frogs, and birds. The insects in this picture never go under the water. They stay on the surface. The duckweed plant floats on the water. It is also food for birds and fish.

How a gnat grows

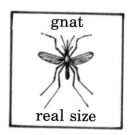

gnat

real size

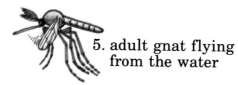

5. adult gnat flying
from the water

4. adult gnat hatching
from pupa

1. raft of eggs

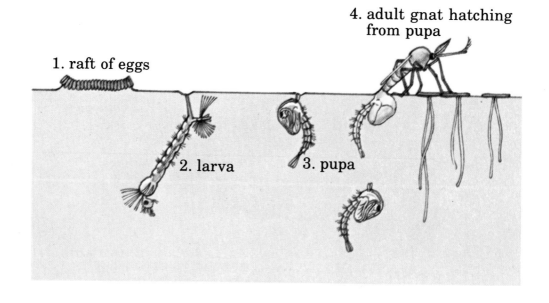

2. larva

3. pupa

Gnats are food for frogs and fish.
Gnats spend most of their life in the
water. Fish eat the eggs, larva, and
pupa. Some gnats become adults. Some
of the adults are eaten by the frogs.

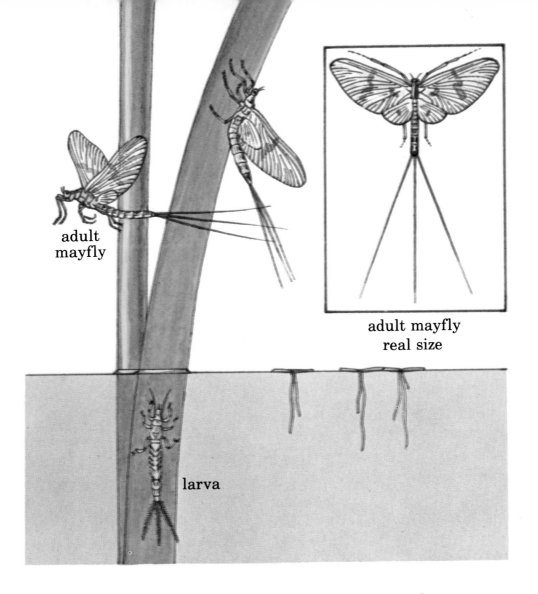

adult
mayfly

adult mayfly
real size

larva

The larva of a mayfly lives for
one or two years in the water. The
adult mayfly has no mouth, so it
cannot eat. Adult mayflies live only a
few hours.

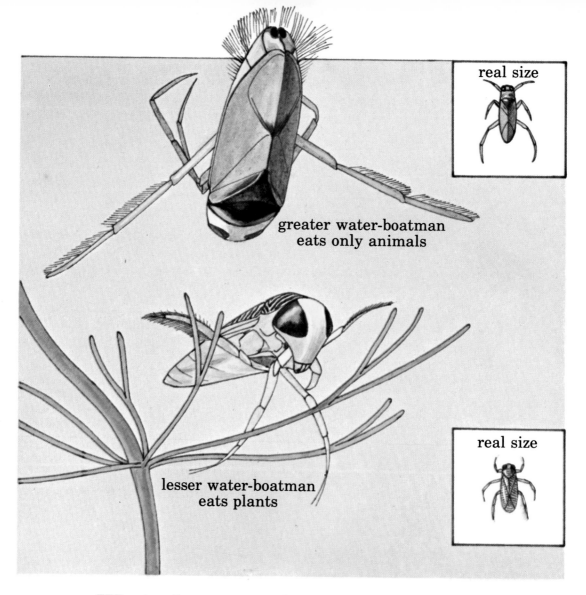

real size

greater water-boatman
eats only animals

lesser water-boatman
eats plants

real size

Water-boatmen have one set of legs
that act like oars. They can swim very
fast. The hairs on the legs push
against the water. This helps them
swim. They come to the surface
for air. One kind of boatman swims
upside down.

male great
diving beetle

female great
diving beetle

This large diving beetle floats with
its head down in the water. It is
trapping air under its wing cases. It
can go underwater and breathe this
air for several minutes.

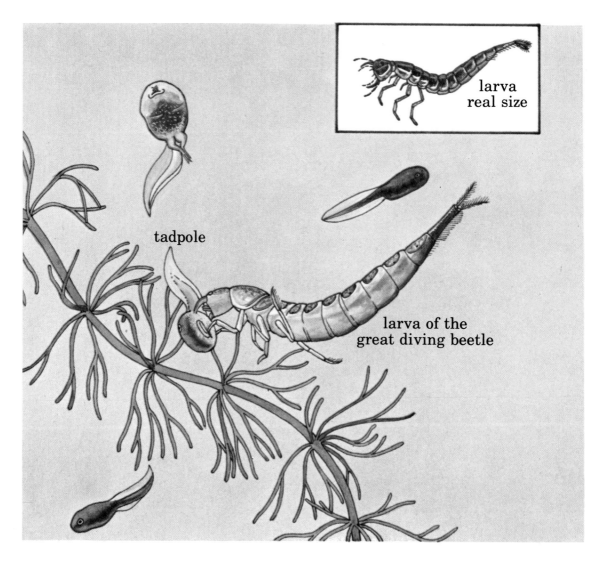

larva
real size

tadpole

larva of the
great diving beetle

The larva of the great diving beetle is very fierce. It attacks and eats many other small pond animals. It comes to the surface for air. It traps air with the hairs around its tail. It can breathe for several minutes with this air.

18

These small plants are from the
pond. The girl sees many small
animals on the plants. Some of them
wash off easily. Other animals hold
on tightly.

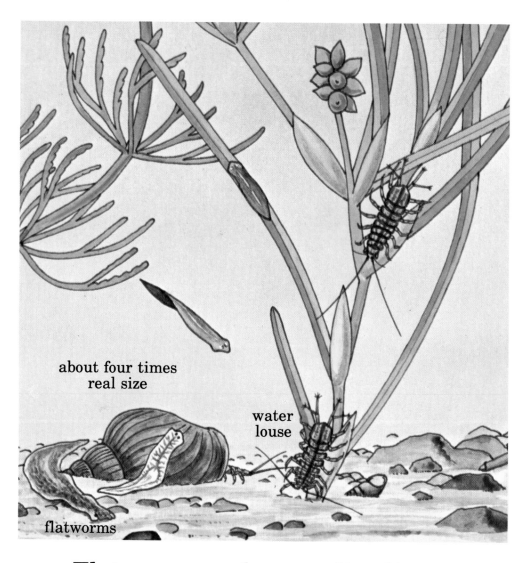

about four times
real size

water
louse

flatworms

Flatworms and water lice live on
the bottom of the pond. They crawl on
the plants. They eat bits of food left
by other animals. We call them
scavengers. The catfish is also
a scavenger.

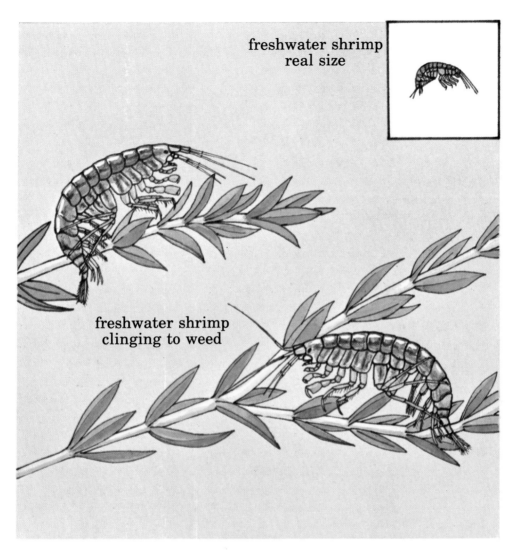

freshwater shrimp
real size

freshwater shrimp
clinging to weed

Freshwater shrimp live in the pond.
They have a hard shell and many
pairs of legs. They live in clear water.
They do not grow big enough for us
to eat.

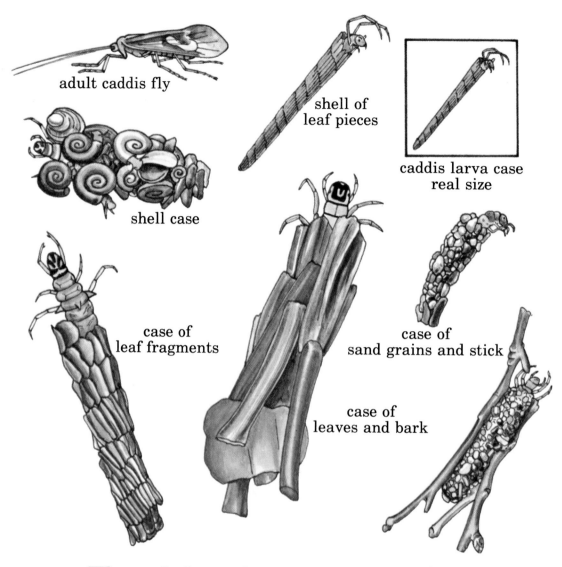

adult caddis fly

shell of
leaf pieces

caddis larva case
real size

shell case

case of
leaf fragments

case of
sand grains and stick

case of
leaves and bark

The adult caddis fly lays eggs on
the shore of the pond. The larva goes
into the pond to live. It then forms a
case around itself. The case can be
made of many different things. The
adult caddis fly hatches from this case.

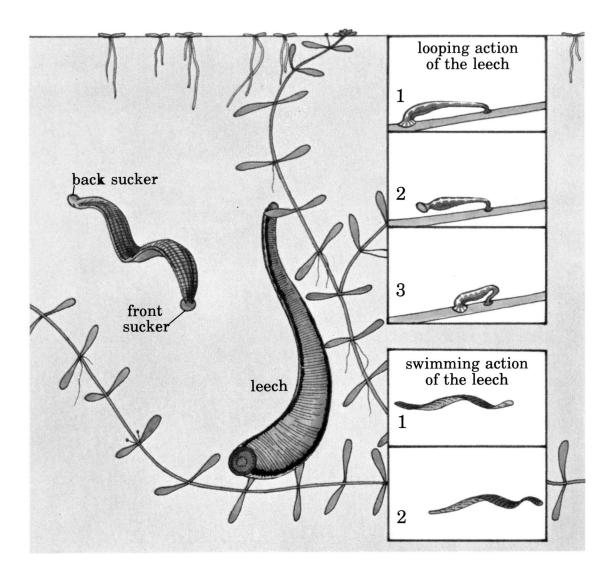

back sucker

front
sucker

leech

looping action
of the leech

1

2

3

swimming action
of the leech

1

2

Leeches look like flatworms. The
leech has a sucker at each end of its
body. It uses the suckers to hold on to
plants and stones. Leeches are good
swimmers. When they swim, their
bodies move like waves.

great ramshorn

freshwater winkle

great pond snail

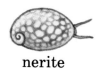

nerite

Jenkin's spire snail

Snails carry their houses on their backs. Their shells are their houses. The shells are very pretty. There are many sizes and shapes of snails in the pond.

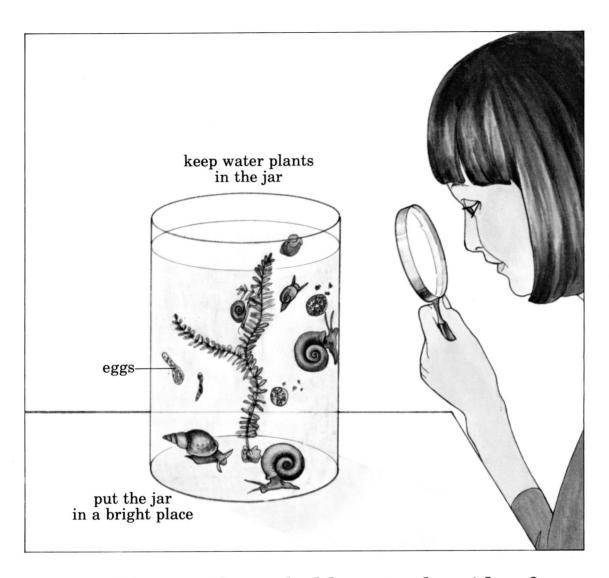

keep water plants
in the jar

eggs—

put the jar
in a bright place

The snail can hold on to the side of the jar. The girl is using a magnifying glass to watch snails hatch from eggs. Snails eat things left behind by other animals. Snails are scavengers.

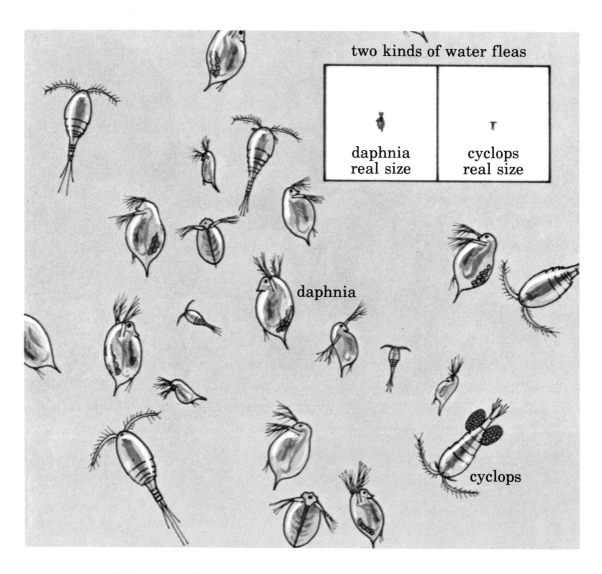

two kinds of water fleas

daphnia real size	cyclops real size

daphnia

cyclops

Water fleas are very small. A magnifying glass helps us to see them better. There are lots of water fleas in the pond. Larger animals eat them for food.

bucket with
pond water

aquarium tank

wide-necked
jar

waterweed

container

The children are going to keep some
pond animals and plants for a short
time. They are being careful so that
the animals will live. They do not
want them to die before they take
them back to the pond.

leeches

stickleback

great diving
beetle

snails, caddis larvae,
water lice

water-boatman

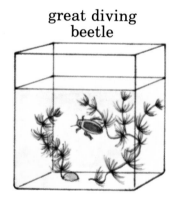

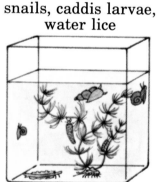

Most of the animals have been put
by themselves. The snails, caddis
larvae, and water lice will not eat each
other. The stickleback needs live water
fleas to eat. The other animals just
need water and some plants for food.

The children watch the animals and
draw pictures. They can make a book
about the pond from the drawings.
After a few days they return the
animals to the pond.

What I Know About Pond Life

The pond is a community.

Plants and animals live in a pond.

The roots of some plants are attached
to the bottom of the pond.

Some water plants have flowers.

Some animals live in the water.

Some animals live on the water.

Other animals live on the bank.

Animals lay eggs on the bank.

Other animals lay eggs in the water
near the bank.

Young frogs and toads are called tadpoles.

Tadpoles have no legs.

Insects in and around the pond are
food for frogs, fish, and birds.

A snail carries its house with it.

The snail has a foot that holds on so
it can climb on things.

Can You Answer These Questions?

1. Do all pond plants attach roots to the bottom of the pond?

2. Is the stickleback a fish or a plant?

3. What does the stickleback eat?

4. Where do frogs lay eggs?

5. What does the tadpole eat?

6. The adult mayfly lives only a few hours. Why?

7. How does the water-boatman swim?

8. What does the leech have on each end of its body?

9. Where does a snail live?

10. Snails eat things left behind by other animals. What do we call animals that do this?

31

Words in POND LIFE

bank
page 5

case
page 22

roots
page 6

sucker
page 23

surface
page 13

magnifying
glass
page 25

bottom
page 20

container
page 27